ANSWERS TO

FIRST AID IN SCIENCE

Robert Sulley

HODDER
EDUCATION
AN HACHETTE UK COMPANY

Hachette UK's policy is to use papers that are natural, renewable and recyclable products and made from wood grown in well-managed forests and other controlled sources. The logging and manufacturing processes are expected to conform to the environmental regulations of the country of origin.

Orders: please contact Hachette UK Distribution, Hely Hutchinson Centre, Milton Road, Didcot, Oxfordshire, OX11 7HH. Telephone: +44 (0)1235 827827. Email education@hachette.co.uk. Lines are open from 9 a.m. to 5 p.m., Monday to Friday. You can also order through our website: www.hoddereducation.com

© Robert Sulley 2013
First published in 2013 by
Hodder Education,
An Hachette UK Company,
Carmelite House
50 Victoria Embankment
London EC4Y 0DZ

Impression number   5
Year   2022

Illustrations by Pantek Media, Barking Dog Art and Robert Hichens Designs
Typeset in Garamond-Book 11 pt by Integra Software Services Pvt. Ltd., Pondicherry, India.
Printed and bound by CPI Group (UK) Ltd, Croydon, CR0 4YY

A catalogue record for this title is available from the British Library

ISBN 978 1 444 18645 1

Contents

Page

Introduction   iv

Living Things   **1**
  1 Life and Living Things   1
  2 Plants   3
  3 The Human Body   5
  4 Nutrition and Health   7

Energy, Forces and Motion   **9**
  5 Forces   9
  6 Energy   11
  7 Light and Sound   13
  8 Electricity   17

The World Around Us   **19**
  9 Rocks, Minerals and Soil   19
 10 Water and Air   21
 11 The Environment   23
 12 The Earth in Space   25

Revision Tests   **27**
Revision Test 1   27
Revision Test 2   30
Revision Test 3   34

Introduction

This book gives the answers to all the exercises and revision tests in *First Aid in Science*. The page numbers from *First Aid in Science* are given next to the answers to help you find the right exercise or revision test.

If you are a teacher or parent, this book will help you check the work of your pupils or children. If you are a pupil, you can use this book to check your own work. Remember not to look at the answers until you have tried to answer the questions.

If you have a wrong answer and do not know why, look back at *First Aid in Science* and read through the explanations again.

Enjoy your Science!

Robert Sulley

Living Things

1   Life and Living Things

Exercise 1 Page 3

There are <u>seven</u> life processes. The life process which means using food to stay alive is called <u>nutrition</u>. To turn food into energy, most plants and animals need to use <u>oxygen</u>. This life process is called <u>respiration</u>. All living things make new living things like themselves. This life process is called <u>reproduction</u>.

Exercise 2 Page 3

Living: frog, goat, seaweed, grass, worm, fern, fish, mosquito, lizard, tree

Not living: eraser, pencil, glass of water, car, fire, rock, chopped wood

Exercise 3 Page 4

Plant kingdom	Animal kingdom	Fungi kingdom	Microbe kingdoms
potato carrot	human earthworm bird locust	toadstool	amoeba

Exercise 4 Page 6

a) A horse is a **vertebrate**.
b) A frog is a **vertebrate**.
c) A worm is an **invertebrate**.
d) An eagle is a **vertebrate**.
e) An ant is an **invertebrate**.
f) A bee is an **invertebrate**.
g) A human is a **vertebrate**.

Revision Test on Life and Living Things Page 7

1. Movement, respiration, growth, reproduction, excretion, nutrition, sensitivity
2. Respiration
3. Reproduction
4. Nutrition
5. Growth
6. The animal kingdom
7. The plant kingdom
8. Dogs and sharks have backbones and so they are vertebrates.
   Moths, worms and spiders do not have backbones and so they are invertebrates.
9. Humans, monkeys and dolphins are all mammals.
10. An arachnid

2 Plants

Exercise 1 Page 9

a) Roots take in water and minerals.
Roots keep the plant in place.
Roots sometimes store food for the plant.

b) Tap roots, fibrous roots and storage roots

c) There are lots of plants with storage roots. Cassava, yams, carrots, beet, sweet potatoes and turnips are all examples of plants with storage roots. You can choose any one to answer this question.

d) Tap roots

Exercise 2 Page 11

Leaves make food for the plant using sunlight. This is called <u>photosynthesis</u>. Leaves lose water through tiny holes in the surface. This draws more water through the plant. This process is called <u>transpiration</u>. Leaves also use the tiny holes to get rid of waste gases. This process is called <u>excretion</u>. The <u>petiole</u> is the stalk attaching the leaf to the <u>stem</u>. The midrib carries <u>water</u> to the leaf and dissolved sugar to and from the leaf.

Exercise 3 Page 13

a) The petals

b) The stigma

c) Pollination and fertilisation

d) Pollination

e) By animals (usually insects or small birds) or by the wind

f) The ovules

Exercise 4 Page 15

a) A seed

b) The ovary

c) By wind, by water, by animals or by explosion

Revision Test on Plants Pages 16–17

1. Water and minerals
2. A fibrous root
3. Leaves make food for the plant by using sunlight to turn water and carbon dioxide into sugar. This process is called photosynthesis.

4.

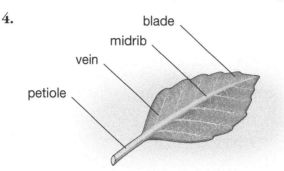

5.

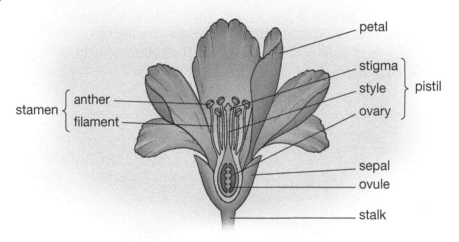

6. The anther
7. The stigma
8. By animals (usually insects or small birds) or by the wind
9. Fertilisation is when the pollen joins together with the <u>ovule,</u> which then grows into a seed.
10. The fruit of the plant
11. Seeds are dispersed by the wind, by water, by animals or by explosion. You could pick any two of these to answer this question.
12. Germination

3 The Human Body

Exercise 1 Page 21

a) Your skeleton protects your body parts, supports the body and allows movement because of the joints between the bones.

b) The ribs

c) The skull

d) Muscles help us move by pulling on the bones to make the skeleton bend at the joints.

Exercise 2 Page 24

a) The heart

b) Arteries, veins and capillaries

c) Capillaries allow gases and food to move into and out of the blood.

d) The arteries

e) Oxygen

f) When you run, your muscles need more energy and so your heart beats faster and you breathe faster to get more food and oxygen to the muscles.

Exercise 3 Page 27

a) The oesophagus

b) The liver produces a liquid called bile which helps to break down the fat in food.

c) In the small intestine

d) It passes through the walls of the small intestine into our blood.

e) In the testes

f) An embryo

Revision Test on The Human Body Page 29

1.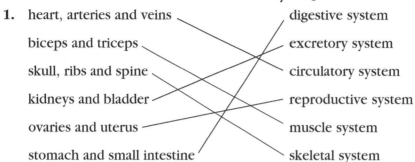

   heart, arteries and veins — circulatory system

   biceps and triceps — muscle system

   skull, ribs and spine — skeletal system

   kidneys and bladder — excretory system

   ovaries and uterus — reproductive system

   stomach and small intestine — digestive system

2. The circulatory system moves blood around our bodies.
3. The veins
4. About 70 beats each minute
5. Oxygen
6. Carbon dioxide
7. Capillaries
8. The liver
9. Mouth → oesophagus → stomach → small intestine → large intestine
10. In the ovaries
11. In the testes
12. About nine months

4 Nutrition and Health

Exercise 1 Page 33

a) Carbohydrates, proteins, fats, vitamins and minerals
b) We must eat fibre and drink water.
c) Carbohydrates and fats
d) We need to eat fibre to help digestion.
e) There are many foods that contain fibre. Common ones are cereals, wholegrain bread, fruit and vegetables. You can choose any two to answer this question.
f) The grease spot test
g) Iodine solution

Exercise 2 Page 36

a) Malnutrition
b) A disease that passes from one person to another
c) Communicable diseases can be spread through the air, in dirty water, through a bite from an insect that has already bitten someone else or by contact with the blood of an infected person. You can choose any two of these to answer this question.
d) HIV/AIDS is spread by contact with the blood of an infected person, usually through unprotected sex or by injecting with a needle that has been used by an infected person.
e) Typhoid, cholera and hepatitis A are spread in dirty water. You can choose any one to answer this question.
f) An injection given by a doctor or a nurse that protects us against diseases.

Exercise 3 Page 38

a) Smoking causes heart attacks, blocked arteries, lung cancer and breathing problems. You can choose any two to answer this question.
b) Nicotine
c) The brain
d) Too much alcohol damages the liver, the heart and the stomach. You can choose any two to answer this question.

Revision Test on Nutrition and Health Page 39

1. Carbohydrates
2. There are many foods that give us protein. Some of the most common are fish, meat, milk, eggs and beans. You can choose any two to answer this question.
3. Rice
4. About 70%
5. The biuret test
6. Some of the most common diseases spread through the air are tuberculosis (TB), chicken pox, measles and influenza ('flu'). You can choose any two to answer this question.
7. Two of the most common diseases that are spread by insect bites are malaria and dengue fever. You can choose any one to answer this question.
8. Lung cancer
9. An epidemic
10. Smoking causes lung cancer and drinking too much alcohol damages the liver.

Energy, Forces and Motion

5   Forces

Exercise 1 Page 42

a) Speed up, slow down, turn, change direction and change shape

b) Lucy and George are pulling in opposite directions on each end of a skipping rope. To start with they don't move so the forces must be <u>balanced</u>. After a while they start to move in the direction George is pulling. The forces must now be <u>unbalanced</u>. George must be pulling with a <u>stronger</u> force than Lucy.

c) Air resistance and friction are two forces that make an object slow down. You can choose one to answer this question.

d) Smooth

Exercise 2 Page 46

a) The pivot

b) There are many examples. Scissors and pliers are two good examples. You can choose any two to answer this question.

c) Magnetism and gravity

d) The magnets will repel each other.

e) A steel nail, an iron horseshoe, a nickel coin

Exercise 3 Page 47

a) Newtons (N)

b) 120 N

c) 120 N

d) 360 N

Revision Test on Forces Page 48

1. Newtons (N)
2. A force meter or a newtonmeter
3. Slow down
4. Without friction the tyres would not grip the road and we would not be able to move forwards.
5. So that the air flows around them easily; this makes the air resistance less.
6. A pivot is the point objects turn around when a force acts on the object.
7. Attract each other
8. Gravity
9. The parachute catches the air which causes a big increase in air resistance. This slows down the parachutist.
10. The force of gravity on the cricket ball is about 1.5 N, so the weight of the cricket ball is about 1.5 N, because 'force of gravity' and 'weight' are the same thing.

6 Energy

Exercise 1 Page 51

a) Kinetic energy
b) Chemical energy
c) Strain energy
d) Gravitational potential energy
e) Chemical energy

Exercise 2 Page 53

a) Chemical energy → electrical energy → light energy + heat energy
b) Chemical energy → heat energy + light energy + a little sound energy
c) Electrical energy → kinetic energy
d) Sound energy → electrical energy → sound energy
e) Gravitational potential energy → kinetic energy

Exercise 3 Page 58

a) The most common non-renewable sources of energy are coal, gas and oil. You can choose any two to answer this question.
b) The most common renewable sources of energy are solar power, wind power and wave power. You can choose any two to answer this question.
c) Heat energy can flow between two things if there is a difference in <u>temperature</u>.
d) Metals are good conductors.
e) Wood and plastic are good insulators. You can choose one to answer this question.
f) Radiation is when heat energy is spread without using particles, such as heat energy from the Sun travelling through space to Earth.

Revision Test on Energy Page 59

1. To answer this question you can choose any five forms of energy from this list: chemical energy, electrical energy, light energy, kinetic energy, gravitational potential energy, strain energy, heat energy. Do not forget to give an example for each one you choose.

2. Chemical energy → electrical energy → sound energy

3. Electrical energy → heat energy + light energy

4. Coal is a non-renewable energy source because it takes millions of years to form from dead plants and animals. So when we use coal it cannot be replaced.

5. Solar power is a renewable energy source because it is replaced every time the Sun shines.

6. A is a liquid; B is a gas; C is a solid.

7. Conduction

8. Convection

9. Convection

10. Metals are good conductors of heat.

11. Wood and plastic are both good heat insulators. You can choose one to answer this question.

12. The handle should be made from a good insulator. If it is made from a good conductor it will burn our hands.

7　Light and Sound

Exercise 1　Page 62

a) A luminous object is an object that gives out light. The Sun, electric lights, torches, candles and fires are all examples of luminous objects. You can choose any one to answer this question.

b) A non-luminous object is an object that does not give out light. Most objects are non-luminous. You can choose any one to answer this question.

c) We can see non-luminous objects when light shines on them. The light bounces off the non-luminous object to our eyes.

d) A transparent material is a material that light can travel through. We can see through transparent materials. Examples are clear glass, clear plastics and water. You can choose any one to answer this question.

e) An opaque material is a material that light cannot travel through, so we cannot see through opaque materials. You can choose any object that we cannot see through to answer this question.

Exercise 2　Pages 65–66

a) When light bounces off an object it is called <u>reflection</u>. When light moves from one transparent substance to another it changes direction slightly. This is called <u>refraction</u>.

b) A mirror and a shiny plate are good reflectors of light.

c)

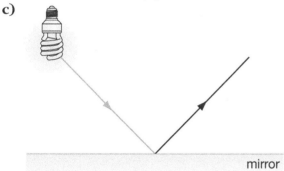

mirror

d) Red, orange, yellow, green, blue, indigo and violet

e) Because it absorbs all the colours in white light apart from blue. The blue light is reflected so the object looks blue.

Exercise 3 Page 68

a) The pupil

b) The retina

c) The optic nerve sends information about what the eye is seeing to the brain.

d) A concave lens

e) A convex lens

Exercise 4 Page 71

a) Sound is caused by an object vibrating.

b) Light

c) The decibel scale

d) A long guitar string

e) Along the auditory nerve from the inner ear

Revision Test on Light and Sound Pages 72–73

1.

2. Opaque materials do not let any light though them, so we cannot see through them.

3. Transparent materials are materials that we can see through. They let light through them.

4. A white wall and a yellow shirt

5. Red, orange, yellow, green, blue, indigo and violet

6. Red light is reflected, all the other colours are absorbed.

7. Refraction

8.

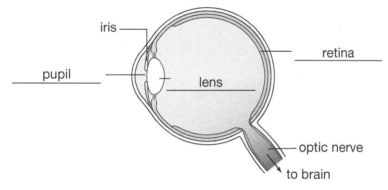

9. Light travels much faster than sound, so we see the lightning before we hear the thunder.

10. The violin, because the strings are shorter

11. Reflection of sound

12.

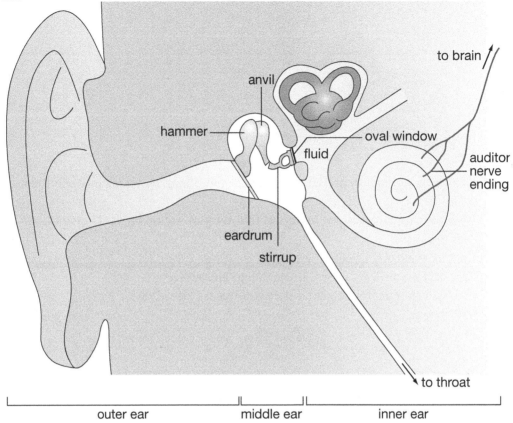

8 Electricity

Exercise 1 Page 77

a) A battery; B bulb; C motor; D switch (open/off)

b) The bulbs in circuits B and C

Exercise 2 Page 78

a) Aluminium and iron

b) Wood, glass and rubber

c) The conductors

Revision Test on Electricity Page 79

1. You can pick any five things that need electricity to work to answer this question. Examples are TVs, refrigerators, electric lights, computers and telephones.

2. Metals are the best conductors of electricity.

3. Plastic is an insulator, so we can pick up the wire without getting an electric shock from the electric current flowing along the wire.

4.

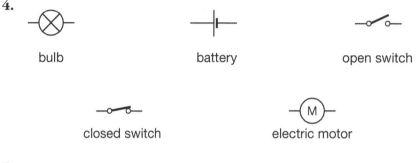

bulb	battery	open switch

closed switch   electric motor

5.

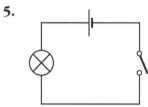

6. It will get brighter.

7. They will be dimmer than a single bulb.

8. The bulbs in circuits A and B

9. Switches are useful because they let us stop the flow of electricity around the circuit so that we can switch things off to save power.

10. Plastic and glass

11. Rubber is a good insulator, so thick rubber gloves will stop the electrician from getting an electric shock.

The World Around Us

9　Rocks, Minerals and Soil

Exercise 1 Page 82

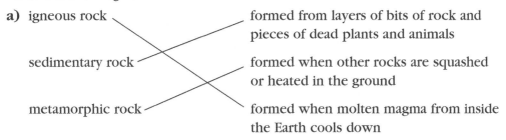

a) igneous rock — formed when molten magma from inside the Earth cools down

sedimentary rock — formed from layers of bits of rock and pieces of dead plants and animals

metamorphic rock — formed when other rocks are squashed or heated in the ground

b) Basalt and <u>granite</u> are examples of <u>igneous</u> rocks. They are very <u>hard</u> rocks. Sedimentary rocks form in <u>layers</u> and are often very <u>soft</u>. Chalk is an example of a soft <u>sedimentary</u> rock. Marble is a <u>metamorphic</u> rock that is often used for monuments and buildings.

c) Igneous rocks are often used for building and for paving roads. Sedimentary rocks are also used for building and some sedimentary rocks are used in the chemical industry. You only need to pick one example for each type of rock to answer the question.

Exercise 2 Page 86

a) Small pieces of rock, humus, air and water
b) Changes in temperature, freezing and thawing of ice, rocks rubbing together and plant roots are the main causes of physical weathering of rocks. You can choose any one to answer this question.
c) Acid in rain water and oxygen in the air and rain water are the main causes of chemical weathering of rocks. You can choose either one to answer this question.
d) The three main kinds of soil are sandy soil, clay soil and loam.
e) The most common ways that people cause soil erosion are cutting down trees, over-grazing and planting the wrong crops. You can choose any two to answer this question.

Revision Test on Rocks, Minerals and Soil Page 87

1. Igneous rocks are made when molten magma that has come from inside the Earth cools down.
2. Common sedimentary rocks are sandstone, limestone, chalk and shale. You can choose any two to answer this question.
3. Chalk
4. Minerals
5. Metamorphic rocks such as marble are often used for monuments and statues. Slate is a metamorphic rock that is used for roof tiles. You can choose one example to answer this question.
6. Soil is made when very small pieces of rock break off from larger rocks.
7. Plant roots grow in the cracks in rocks. As the roots get bigger, they make very small pieces of the rock break off.
8. Small amounts of carbon dioxide dissolve in rain water which makes rain water slightly acidic. The acid reacts with the rocks and makes very small pieces break off.
9. There are not many minerals and not much water in sandy soil, so it is not good for plants.
10. Soil erosion
11. Tree roots help to hold soil together and the leaves that drop from trees form humus, so it is bad for the soil if trees are cut down.
12. People can protect the soil by planting trees on slopes and hillsides, not having too many animals to prevent over-grazing, planting the right crops for the soil, planting different crops each year and terracing land on steep hills. You can choose any two to answer this question.

10 Water and Air

Exercise 1 Page 89

a) 70%

b)
solid — water vapour
liquid — ice
gas — water

(solid links to ice, liquid links to water, gas links to water vapour)

c) 100 °C

d) Water near the surface of the puddle gets enough energy from the Sun to evaporate from the surface and form water vapour.

Exercise 2 Page 92

a) Water in the sea <u>evaporates</u> when the Sun shines on it and forms <u>water vapour</u>. This water vapour rises up and <u>condenses</u> into droplets, forming clouds. When there is a lot of water in the clouds it falls as rain. If the air is very cold it falls as snow. When water falls from the clouds as rain or snow it is called <u>precipitation</u>. The rain water that falls on land runs into rivers and then back into the <u>ocean</u>, where the cycle starts again.

b) Salt water

c) Water can become polluted if people use rivers, lakes or the sea as a toilet or for emptying toilet waste. Water can also become polluted if people or factories dump waste, garbage and rubbish into rivers or the sea, or if waste gets washed into rivers. Pesticides and other chemicals from farms cause pollution if they get washed into rivers. Ships sometimes pollute water if they dump waste and oil into rivers and seas. You can choose any three to answer this question.

d) The most common diseases carried by polluted water are cholera, typhoid, dysentery, hepatitis and poliomyelitis. You can choose any two to answer this question.

e) When water is boiled it kills germs in the water.

Exercise 3 Page 94

a) Nitrogen, oxygen and carbon dioxide

b) The atmosphere

c) Some of the most common causes of air pollution are fumes from cars and trucks, chemicals and smoke from factories, smoke from burning rubbish or garbage, smoke from burning coal and smoke from cigarettes. You can choose any three to answer this question.

d) Two common diseases caused by air pollution are asthma and lung cancer.

Revision Test on Water and Air Pages 95–96

1. 70%
2. Freezes
3. Gas
4. Tiny droplets of water
5. Precipitation is when water falls from clouds as rain or snow.
6. Salt water
7. Polluted
8. Chlorination
9.

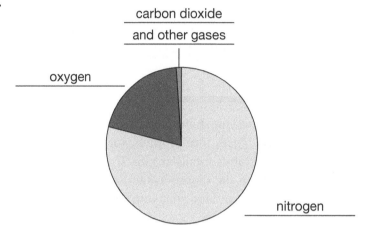

10. Oxygen
11. Atmosphere
12. Air pressure

11 The Environment

Exercise 1 Page 98

a) Habitat

b) Humans can live in lots of different kinds of habitat because we can wear clothes to keep us warm or cool, we can eat lots of different types of food and we can build houses to protect us from the weather. You can choose any two to answer this question.

c) Cactus plants have long roots to find water, fleshy stems that store water and thin needles to slow down water loss. You can choose any one of these adaptations to answer this question.

d) Seals have layers of fat to keep them warm, small ears to reduce heat loss and streamlined bodies to help with swimming. You can choose any one of these adaptations to answer this question.

e) Some examples of adaptations that fish have for living in the sea are streamlined bodies and fins to help them swim, the ability to live in salt water and the ability to get oxygen from the water for respiration.

Exercise 2 Page 99

a) Sulphur and nitrogen

b) Acid rain can kill plants and pollute soil.

c) Greenhouse gases

d) By stopping heat escaping from the atmosphere back into space

Exercise 3 Page 101

a) Two

b) The hawk

c) If there is less grain, the number of mice will go down because they will not have enough to eat. If the number of mice goes down then the number of hawks will also go down as they will also have less to eat.

Revision Test on The Environment Page 103

1. Camels have broad flat feet and long legs so that they do not sink into the sand. They have long eyelashes and nostrils that can close so that they do not get sand in their eyes or nose. They have humps that can store fat for food, and their bodies can store water so that they can go for a long time without eating or drinking. You can choose any two of these adaptations to answer this question.

2. Polar bears have a thick layer of fat to keep them warm. They have thick fur which traps air, also keeping them warm. They have small ears so that they do not lose too much heat through them. All of these adaptations help them to live in a cold habitat. You can choose any one to answer this question.

3. Owls have eyes that have adapted to see at night. They have soft feathers so that they can fly quietly. They also have very good hearing which helps them to find small animals to catch and eat. These adaptations help an owl to hunt at night. You can choose any one to answer this question.

4. Greenhouse gases are gases that make the Earth's atmosphere warmer.

5. Greenhouse gases make the atmosphere warmer because they trap heat around the Earth and stop it escaping into space.

6. Greenhouse gases and acid rain are bad for the environment because they change or destroy habitats so that some plants and animals cannot live.

7. a) The leaves
   b) Three
   c) The centipede or the snake
   d) The number of snakes would go down.

8. The number of krill would go up because there would be fewer whales to eat them.

9. If grassland is destroyed there will not be as many mice and other small animals that owls need to eat. So the owls will start to die.

10. An endangered species is a kind of animal or plant that is dying out.

12 The Earth in Space

Exercise 1 Page 106

a) Jupiter

b) 141 920 km

c) 12 682 km

d) 150 million km

e) 78 million km (because the average distance of Mars from the Sun is 228 million km)

f) 58 million km

g) 88 Earth days

Exercise 2 Page 108

a) 24 hours

b) Because the other side of the Earth is facing away from the Sun and so does not get any light.

c) Summer

d) 28 days

e) Because we cannot always see all of the Moon. We can only see those parts of the Moon that reflect light from the Sun towards Earth.

f) Because it takes the Moon exactly the same time to spin once on its axis as it takes for it to orbit the Earth, so the same side always faces Earth.

Exercise 3 Page 110

a) A galaxy is a large group of stars.

b) The Milky Way

c) A spiral

d) About 2500

Revision Test on The Earth in Space Page 111

1. Spheres
2. The Earth orbits around the Sun. It makes one orbit every 365.25 days.
3. Mars, Earth, Jupiter
4. A day
5. Winter
6. Because when it is summer at the South Pole the southern hemisphere is pointing towards the Sun. So the northern hemisphere is pointing away from the Sun and it is winter at the North Pole.
7. The Earth
8. None of it
9. A galaxy
10. Just over four years

Revision Tests

Revision Test 1 Pages 112–115

1. Movement, respiration, growth, reproduction, excretion, nutrition, sensitivity
2. Reproduction
3. The leaves
4. Seeds and fruits are dispersed by wind, water, animals and explosion.
   You can choose any three to answer this question.
5. Arteries, veins and capillaries
6. The liver
7. The iodine test
8. Speed up, slow down, change direction, turn around a point or change shape
9. Air resistance and friction are both forces that slow down a moving object.
   You can choose either to answer this question.
10. Attract
11. Newtons
12. Red, orange, yellow, green, blue, indigo and violet
13. The lens
14. Light
15. An open switch
16. Igneous, sedimentary and metamorphic
17. There are not many minerals in sandy soil and not much water, so plants do not grow well.
18. Tree roots help to hold soil together so if trees are cut down it is easier for water to wash away the soil. Also, leaves from the trees drop to the ground and form humus which is needed for good soil.
19. Ice
20. X
21. Oxygen
22. Carbon dioxide is the main greenhouse gas.
23. The leaves
24. Winter
25. 24 hours
26. Which of these four animals is not a mammal?
    A cat
    Ⓑ snake
    C cow
    D mouse

27. In this drawing, the part of the plant labelled X is:
    Ⓐ the ovary
    B the pistil
    C the stamen
    D the petal

28. Which of these is the way that water reaches the leaves of plants?
    A stem → roots → leaves
    Ⓑ roots → stem → leaves
    C leaves → stem → roots
    D roots → leaves → stem

29. Which of these is not an organ of the human body?
    A lung
    B heart
    Ⓒ finger
    D liver

30. Where does most digestion take place in the human body?
    Ⓐ small intestine
    B stomach
    C large intestine
    D oesophagus

31. Which of these nutrients gives us lots of energy?
    A proteins
    B minerals
    C vitamins
    Ⓓ carbohydrates

32. Which of these materials is a good electrical insulator?
    A copper
    B aluminium
    Ⓒ plastic
    D steel

33. How much of the human body is made up of water?
    A about 15%
    B about 50%
    C about 40%
    Ⓓ about 70%

34. Which one of these kinds of energy is stored in a stretched elastic band?
    A chemical energy
    Ⓑ strain energy
    C kinetic energy
    D gravitational potential energy

35. When heat spreads through a gas from hot to cold it is called:
    Ⓐ convection
    B radiation
    C insulation
    D conduction

36. Which of these statements is true?
    A Sedimentary rocks are made when lava from a volcano cools.
    B Sedimentary rocks are made when other rocks are heated inside the Earth.
    Ⓒ Sedimentary rocks are made when bits of rock and dead plants and animals form layers.
    D Sedimentary rocks are made when other rocks are weathered by the wind and rain.

37. At what temperature does water freeze?
    A 50 °C
    B −100 °C
    Ⓒ 0 °C
    D 100 °C

38. Which gas makes up about 78% of the air around us?
    A oxygen
    B carbon dioxide
    C hydrogen
    Ⓓ nitrogen

39. The Sun and all the things that orbit around it together are known as:
    A the Universe
    B the Galaxy
    C the Milky Way
    Ⓓ the Solar System

40. Which of these is the name given to the imaginary line around the middle of the Earth?
    Ⓐ the equator
    B the hemisphere
    C the orbit
    D the Arctic circle

Revision Test 2   Pages 116–120

1. Nutrition
2. The animal kingdom
3. C
4. Plants can be pollinated by insects, birds or other animals, or by the wind.
5. Petiole
6. Arteries
7. The most common foods that give us protein are fish, meat, milk, eggs and beans. You can choose any one to answer this question.
8. Typhoid, cholera and hepatitis A are all spread in dirty water. You can choose any one to answer this question.
9. Magnetism and gravity are both examples of non-contact forces. You can choose either one to answer this question.
10. 500 N
11. B
12. Any object that gives out light is luminous, such as an electric light.
13. We see lightning before we hear thunder because light travels much faster than sound.
14.

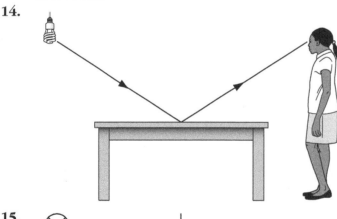

15.

16. Igneous rocks are made when molten magma from inside the Earth cools down.
17. Water can become polluted if people use rivers, lakes or the sea as a toilet or for emptying toilet waste. Water can also become polluted if people or factories dump waste, garbage and rubbish into rivers or the sea, or if waste gets washed into rivers. Pesticides and other chemicals from farms cause pollution if they get washed into rivers. Ships sometimes pollute water if they dump waste and oil into rivers and seas. You can choose any two ways to answer this question.
18. Boiling, filtration and chlorination are three ways of purifying water.

19. About 3%

20. The atmosphere is the layer of air around the Earth.

21. Greenhouse gases make the Earth warmer by trapping heat in the atmosphere so that it cannot escape into space.

22. Cactus plants have long roots to find water, fleshy stems that store water and thin needles instead of big leaves so that they lose less water through their leaves. All of these adaptations help a cactus to live in a hot and dry habitat. You can choose any two to answer this question.

23. 365.25 days

24. We always see the same side of the Moon because it spins once on its axis in exactly the same time as it takes it to make one complete orbit of the Earth. So the same side is always facing the Earth.

25. The Milky Way

26. Which of these is a vertebrate?
   Ⓐ an animal with a backbone
   B an animal without a backbone
   C a flowering plant
   D a bacterium

27. Which of these is the sticky part of the flower that collects pollen?
   A ovary
   B anther
   Ⓒ stigma
   D petal

28. Which gas does blood give to the air in the lungs to be breathed out?
   A oxygen
   Ⓑ carbon dioxide
   C nitrogen
   D hydrogen

29. Which of these foods is a good source of carbohydrate?
   A beans
   B carrots
   Ⓒ bread
   D tomatoes

30. In this tug-of-war game, no one is moving.

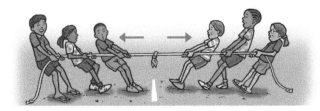

How would you describe the forces acting on the rope?
   A unbalanced forces   Ⓒ balanced forces
   B pushing forces   **D** turning forces

31. A force can make an object turn around a point. What name is given to the point an object turns around?
    A an axis
    B an angle
    C a lever
    D a pivot

32. Which of these is the energy of movement?
    A chemical energy
    B kinetic energy
    C gravitational potential energy
    D electrical energy

33. Which of these materials is a good conductor of heat?
    A plastic
    B wood
    C metal
    D rubber

34. Heat from the Sun gets to the Earth by:
    A convection
    B radiation
    C conduction
    D reflection

35. Which of these materials is a good electrical insulator?
    A rubber
    B aluminium
    C iron
    D copper

36. A farmer is looking at the soil. Which soil will be best for growing crops?
    A sandy
    B clay
    C loam
    D gravel

37. When liquid water changes to gas at 100 °C, we say that the water:
    A freezes
    B condenses
    C melts
    D boils

38. Here is a drawing of the Earth orbiting the Sun.

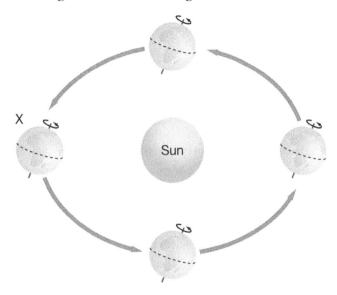

When the Earth is at point X, it is:

A winter in the northern hemisphere and summer in the southern hemisphere

B summer all over the Earth

Ⓒ summer in the northern hemisphere and winter in the southern hemisphere

D autumn in the northern hemisphere and spring in the southern hemisphere

39. Which is the largest planet in the Solar System?

A Mercury

Ⓑ Jupiter

C Earth

D Mars

40. Which of these is a galaxy?

A the Sun and all the things that move around it

B the orbit of the Moon around the Earth

Ⓒ billions of stars grouped together

D the distance from the nearest star to Earth

Revision Test 3 Pages 121–124

1. Respiration
2. Growth
3. The animal kingdom
4. Plant roots take in water and minerals from the soil, they keep the plant in place, and some plant roots store food.
5. Nine months
6. Carbohydrates, proteins, fats, vitamins, minerals
7. Communicable diseases can be spread through the air, in water, by insect bites and sometimes directly from the person who is ill. You can choose any two to answer this question.
8. Gravity
9. The magnets will repel each other.
10. It takes millions of years for dead plants and animals to be changed into oil, so when we use up oil we cannot renew it.
11. Solar power is renewable because it is renewed when the Sun shines.
12. The eardrum
13. Refraction
14. An echo
15. Decibels
16. Circuits B and C
17. Metamorphic rocks are made when other rocks are squashed or heated in the Earth.
18. People cause soil erosion by cutting down trees, keeping too many animals on the land which causes over-grazing and by planting the wrong crops. You can choose any two to answer this question.
19. Physical weathering of rocks is caused by changes in temperature, freezing and thawing of ice, rocks rubbing together and plant roots growing in cracks. You can choose any two to answer this question.
20. The three main gases in air are nitrogen, oxygen and carbon dioxide.
21. Air pressure
22. Polar bears have a thick layer of fat to keep them warm. They have thick fur which also keeps them warm. They have small ears so that they do not lose too much heat through them. You can choose any two to answer this question.
23. The mouse and the hawk
24. The Earth
25. One day
26. Which of these is *not* one of the seven life processes?
    A movement
    Ⓑ hearing
    C sensitivity
    D nutrition

27. Which of these is the part of the flower that makes pollen?
    A stalk
    B sepals
    Ⓒ anther
    D ovary
28. When plant seeds start to grow it is called:
    Ⓐ germination
    B reproduction
    C dispersal
    D pollination
29. Which of these is *not* a job done by your skeleton?
    A It protects your body parts.
    Ⓑ It helps you digest food.
    C It supports your body.
    D It lets you move.
30. Which of these is a test to see if a food has fat in it?
    A the iodine test
    B Benedict's test
    Ⓒ the grease spot test
    D the biuret test
31. Which one of these things will be attracted to a magnet?
    A a wooden pencil
    B a can made of aluminium
    C a plastic spoon
    Ⓓ a steel nail
32. Which one of these instruments can be used to measure force?
    Ⓐ a newtonmeter
    B a voltmeter
    C a thermometer
    D an ammeter
33. Which of these types of energy is made by a vibrating object?
    A chemical energy
    B electrical energy
    Ⓒ sound energy
    D light energy

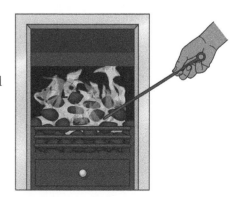

34. When heat from a fire travels along a metal poker it is called:
    A convection
    Ⓑ conduction
    C radiation
    D insulation

35. What name do we give to a material that doesn't allow light to pass through it?

   A transparent

   Ⓑ opaque

   C luminous

   D translucent

36. Which of these materials is a good electrical conductor?

   Ⓐ iron

   B plastic

   C wood

   D rubber

37. When water changes from a gas to a liquid we say the water:

   A freezes

   B evaporates

   Ⓒ condenses

   D melts

38. At what temperature does water boil?

   A 0 °C

   Ⓑ 100 °C

   C −100 °C

   D 20 °C

39. Which of these causes acid rain?

   Ⓐ gases from pollution dissolved in rain water

   B rubbish and garbage thrown into rivers

   C physical weathering of rocks

   D oil pollution in the sea evaporating to make rain

40. What is the biggest object in the Solar System?

   A Earth

   B Jupiter

   Ⓒ the Sun

   D Saturn